*Geology of the
Eastern Snake River Plain Field Trip:
Part II Craters of the Moon National
Monument and the Great Rift*

By William A. Szary

Copyright 2021. Earth2Energy. All Rights Reserved.

Entrance trail leading to Crystal Ice Caves, Idaho near Kings Bowl. Layered basalts are covered by sedimentary deposits. Source: Flikr posted on the internet.

Book Cover: Front: Bigs Craters located in Craters of the Moon National Monument are aligned along the Great Rift. Source: The Earth Story posted on the internet. Back: The King's Bowl fissure features the site of southern Great Rift eruptions. Source: Google Earth.

Library of Congress Catalog in Publications Data:

Szary, William A. Geology of the Eastern Snake River Plain Field Trip: Part II Craters of the Moon National Monument and the Great Rift
Includes references

ISBN 13: 9798461763510

Earth2Energy Educational Publishing
Port Richey FL 34668

Earth2Energy is a Registered Trademark

Table of Contents

Field Excursion 1. Eastern Snake River Plain Field Trip: Craters of the Moon National Monument and the Great Rift 4
Radiocarbon Dating and Evolution of the Craters of the Moon Lava Field
Paleomagnetic Studies
Petrography and Chemical Composition of Crater of the Moon Lavas
Lava Volumes

Field Trip Road Log
Idaho Falls to Craters of the Moon National Monument to Pocatello
Stop 1. Menan Volcanic Complex
Stop 2. Hells Half Acre
Stop 3. Box Canyon Graben
Stop 4. Craters of the Moon Lava Field
Stop 5. Craters of the Moon National Monument
Stop 6. Cerro Grande Volcanic Field
Stop 7. Cedar Butte
Stop 8. North Robbers Lava Field

Field Excursion 2. The Southern Great Rift 49
Kings Bowl Lava Field
Wapi Lava Field

Bimodal Magmatism, Basaltic Volcanism, Tectonics, and Geomorphic Processes in the Eastern Snake River Plain
Geological Overview

Field Trip Guide
Stop 1. Kings and Queens Bowls
Stop 2. Kings Bowl Fissures (Optional)
Stop 3. Wapi Park Lava Field
Stop 4. Split Butte

Photo Gallery of the Eastern Snake River Plain and the Great Rift 64

References 70

Field Excursion 1. Eastern Snake River Plain Field Trip: Craters of the Moon National Monument and the Great Rift

Kuntz and others (1983) arranged a field trip viewing Holocene basaltic volcanism along the Great Rift summarized in this chapter.

Radiocarbon Dating and Evolution of the COM Lava Field

Early studies of the COM (Craters of the Moon) lava field outlined the general stratigraphic relations among lava flows. Later studies refined previously established stratigraphy, produced detailed geologic maps of the lava field, and attempted to date key flows by the radiocarbon method. Samples for radiocarbon studies include charcoal from tree molds and organic material in sediment buried by flows. The radiocarbon and stratigraphic data indicate that the COM lava field formed during at least eight eruptive periods (H, the oldest, through A, the youngest) that began about 15,000 and ended about 2,100 yrs ago. Each eruptive period was probably only a few tens of years or possibly a few hundred years long. Eruptive periods were separated by intervals of quiescence that lasted a few hundred to a few thousand years.

Only one or a few flows in each eruptive period have been dated. Thus, the duration of each eruptive period is not established, and the evolution of the COM lava field through the eight eruptive periods is indeterminant. Individual lava flows are described along with their source vents along with the approximate ages of each eruptive period. Volcanic features to be viewed on the first day of the trip are chiefly those of the youngest eruptive period (A) of the COM lava field.

Paleomagnetic Studies

Paleomagnetic measurements have been used to correlate and approximately date lava flows and groups of flows to aid in deciphering the volcanic history of the lava fields along the Great Rift. Lava flows record the local geomagnetic field at the time of eruption and cooling. Geomagnetic field changes due to secular variation occur at a geologically rapid rate (4°/century) and thus permit assignment of lava flows to groups that have similar paleomagnetic direction. Conversely, dissimilar paleomagnetic directions suggest that two lava flows do not belong to the same group.

Paleomagnetic studies have helped to solve difficult stratigraphic problems for the COM lava field. For example, it was initially thought that all thick aa flows in the southern part of the COMNM (NM- National Monument) were erupted in the same eruptive period from a source vent at or near Big Cinder Butte.

Paleomagnetic inclination data indicate two distinct eruptions of aa lava flows from at least two source vents occurred. Thus, a recognized group of aa flows in eruptive periods C and D from source vents at Big Cinder Butte and Silent Cone are present. Petrochemical data and radiocarbon ages led to split one of these two aa groups into two additional groups. Initially, the Lava Point aa flows in eruptive period D are now placed in period E. Because they have identical paleomagnetic field directions, they are correlated with the Sunset and Carey flows that moved northeast and southwest from the northern part of the Great Rift. Without the paleomagnetic data, it was previously assumed that the Carey and Sunset flows represented separate eruptions from separate source vents.

Petrography and Chemical Composition of COM Lavas

Snake River Plain basalts are texturally, mineralogically, and chemically uniform. The basalts are olivine tholeiites consisting typically of olivine (Fo80), plagioclase (An70-50), ferroaugite, titanomagnetite, ilmenite, and brown glass. An average chemical composition from 37 analyses was used for the determination. COM flows differ significantly from typical Snake River Plain basalt. COM lavas contain phenocrysts of olivine (Fo50-10), plagioclase (An60-40), brown clinopyroxene, titanomagnetite, ilmenite, and brown glass.

Some contain orthopyroxene and apatite. Evolved basalts (i.e., greater than about 52 percent SiO_2) contain xenoliths and xenocrysts of corroded anorthoclase, plagioclase (An55-1S), green clinopyroxene, olivine (Fo25-10) and zircon. Average chemical analyses of several different types of COM flows illustrate the degree of chemical variability. These flows range from alkali basalt to latite. Note that the most primitive COM flows differ from the average Snake River Plain olivine tholeiite in that they have more TiO_2, total iron, Na_2O, K_2O, and P_2O_5, and less MgO and CaO. Evolved COM flows first were erupted about 6,600 yrs ago in eruptive period D and continued to be erupted along with more primitive flows in eruptive periods C, B, and A.

Lava Volumes

Stratigraphic relations, areal extent, thicknesses, and radiocarbon ages of COM lava flows provide for volume data. The age and volume data reveal that volcanism along the northern part of the Great Rift is volume predictable, i.e. the volume of each eruption is a function of the elapsed time between it and the preceding eruption. This relationship has implications to the regional stress field and the magma plumbing system. There is a marked change in composition of COM lavas beginning at eruptive period D. This is attributed to the increased rate of volcanism since 6,600 yrs ago to the addition of supplies of evolved (SiO_2 less than 52 percent) lava to the nearly constant rate supply of non-evolved ($SiO_2 > 52$ percent) lava over the last 15,000 yrs.

Field Trip Road Log

Idaho Falls to Craters of the Moon National Monument (COMNM) to Pocatello

Stop 1 is the Menan Buttes north of Idaho Falls. Take Interstate 15 North to Roberts Idaho exit which is Idaho Highway 48 East. Travel to Menan, Idaho. At the east part of town, turn left on N 3600 E Road and travel north crossing the Snake River. The south butte is on the right, and the north butte is straight ahead. N 3600 E Road is also called Twin Buttes Road. North of East Butte Road on the west flank of the north butte is the North Menan Butte Trail. Park here (**Figure 1**).

Figure 1. North and South Menan Buttes are accessed from Twin Buttes Road or N 3600 E from the eastern side of Menan, Idaho.

Menan Volcanic Complex

The Menan Buttes are known to many geologists because they have been used in topographic map exercises of popular introductory laboratory manuals. They are outstanding morphological examples of tuff cones. A few miles south of the Menan Buttes are the remnants of two tuff rings, the North and South Little Buttes which are much smaller and have lower slope angles than the two tuff rings. The deposits are strongly cyclic and contain abundant accidental quartzite pebbles derived from the underlying permeable alluvium. The tuff rings are dominated by two types of bed sets: black, planar-bedded, coarse scoriaceous lapilli tuff bed sets (mostly fall ash with minor surge deposits) were formed during relatively "dry vent" violent strombolian eruptions driven by rapid vesiculation.

Intercalated bed sets of tan cross-stratified fine palagonite tuffs (fall and surge deposits) were formed during relatively "wet vent" surtseyan eruptions driven by phreato-magmatic processes.

The Menan Buttes tuff cones are the largest features of the Menan volcanic complex with reconstructed volumes of 0.7 (North) and 0.3 (South) km3. Dense basalt volume equivalents are 0.4 and 0.2 km3, respectively. They are among the largest terrestrial tuff cones with volumes comparable to those of Diamond Head, Oahu, and Surtse, Iceland. Deposits of the North and South Menan Buttes tuff cones are monotonous, massive to thin-bedded, tan, lithified palagonite lapilli tuff with minor xenoliths of dense basalt and rounded quartzite pebbles. At this stop observe weakly convoluted thin bedded tuff and bed forms that are nearly obliterated by palagonitization.

Return back to Idaho 48 and turn right back to Interstate 15 in the town of Roberts, Idaho. Take Interstate 15 south to the exit at US 20 in Idaho Falls. Proceed west on U.S. 20. Road crosses loess and alluvium-covered Snake River Plain basalt with only local exposures of the flows. Passing the Osgood Road intersection continue on U.S. 20. Passing through the Shelley-New Sweden Road intersection continue on U.S. 20 over irregular topography of hummocky flow surfaces mantled by loess. Skyline Gun Club is on the right. Road begins ascent of Rifle Range Butte flows (**Figure 2**).

One of several vents in the Rifle Range Butte vent complex is on the right (**Figure 3**). The highway leaves the surface of Rifle Range Butte flows and continues west onto the older Butterfly Butte flows (drainage follows contact). More subdued topography of Butterfly Butte flows is due in part to thicker loess rather than younger Rifle Range Butte flows (**Figure 4**). At the Brunt Road intersection continue west on U.S. 20. North Butte is a low shield volcano at about the 4 o'clock position.

At the intersection with an unnamed road just over the crest of a small hill, features in a counterclockwise direction include: Kettle Butte, a low shield volcano; south end of the Lemhi Range in far distance; East Butte, a rhyolite dome. The following features are not shown in the imagery but are visible from this viewpoint on US 20: Big Southern Butte, a rhyolite dome; highest point marks an extremely broad low shield volcano that constitutes the Hells Half Acre lava field.

Figure 2. Loess covers basalt flows on the flat landscape. The gun range is in the center of this image. Basalt flows are exposed in the background left side of the image.

The lava field is the dark sage and cedar-covered area on the horizon south of the highway (**Figure 5**).

Past Rifle Range Butte, US 20 passes the northern end of the Hells Half Acre lava flow (*Stop 2*). There is an unimproved dirt road off to the left and the highway passes the northern tip of the lava flow closest to the highway. Take the dirt road access south and take the right fork which travels along the edge of the lava flow (**Figure 6**). A four wheel high clearance vehicle is recommended for this road. Caution should be taken during the wet season months.

Stop 2. Hells Half Acre

The vent area of the Hells Half Acre lava field is located in the northwest part of the lava field, and lies only 3 miles southeast of this stop. It is a northwest-elongated crater containing several deep collapse pits. A northwest trending eruptive fissure located just south of the crater and short sections of eruptive fissures located between the crater and this field trip stop suggest that the feeder system for the lava field is a northwest-trending dike.

Figure 3. The Rifle Range Butte vent is off to the right traveling west on US 20 past the gun range. A small exposure is present alongside the highway covered with loess at the surface.

Figure 4. A small road cut along US 20 exposes the Rifle Butte basalt past the butte vent shown in Figure 3.

Figure 5. At the crest of the hill at the intersection of the unnamed road mentioned in the text, the Lemhi Range can be see off in the distance on the far right. Kettle butte is to the right of the range. On the left side, the East butte can be viewed from US 20.

Figure 6. Hells Half Acre lava field is accessed from US 20 west of Rifle Range Butte.

This is consistent with the observation that ESRP volcanic rift zones trend northwesterly almost perpendicular to the trend of the plain itself, and to the interpretation that dike orientation in the ESRP is controlled by the same northeast trending extension stress field that produces Basin and Range faulting adjacent to the ESRP.

At this stop, two sets of non-eruptive fissures flanking the eruptive fissure extend northwestwardly from beneath the edge of the Hells Half Acre volcanic field. A several meter thick loess blanket that underlies the lava field obscures much of the fissuring, but in several places the loess has collapsed into the fissures and/or has been eroded into the fissures by percolating or flowing water. The fissure sets separated by a distance of about 1 mile were caused by a zone of extensional stress above the non-eruptive part of the dike and are typical of surface deformation associated with shallow dike intrusion in volcanic rift zones. Fissure widths are generally less than 1 meter, and the fissure walls are very irregular because pre-existing columnar jointing in the lava flows controlled the near surface fissure shape. Parallel sets of non-eruptive fissures are also observed along the Great Rift between Kings Bowl lava field and the Craters of the Moon lava field, and in the Arco volcanic rift zone. In addition, a small graben about 300 meters across with 10 meters of vertical displacement several km long occurs in the northern part of the Arco volcanic rift zone. A visit to this graben will be forthcoming.

The knobs to the south of the apex of the shield are spatter cones along the Hells Half Acre eruptive fissure system. This fissure system is represented by a set of open cracks that extends northwest from the northwest edge of the lava field (**Figure 7**). The road to left leads to Seventeenmile Cave, an opening (skylight) in a lava tube in the Butterfly Butte flow. Just beyond the road to the cave, the highway crosses the contact onto flows from the vent of Kettle Butte (**Figure 8**).

At Mile marker 288, the road to the right leads to a skylight entrance about 4 mi north of highway in a lava tube in Kettle Butte flow. The area is known as "Owl Cave," or the Wasden archeological site. Faunal materials include bison, mammoth, camel, pronghorn, grizzly bear, wolf, and other animals. Fluted points and bone tools suggest that the rock shelter was used by bison and mammoth hunters as long ago as 12,500 yrs B.P.

Figure 7. Hells Half Acres lava field as seen from US 20 on the left side of the highway while traveling west.

Figure 8. Seventeen Mile Cave is on the left side of US 20. A dirt road leads to the cave (red marker). The contact between Butterfly Butte and Kettle Butte flows is believed to be at the change in grade along the highway at the dashed line. A road cut into the lava flow appears just beyond the contact. The rubbly nature of the rocks appears to be fragments scraped off the road during construction, not a true exposure. Kettle Butte is off to the northeast of the cave entrance. On the horizon, the right volcanic structure is Big Southern Butte.

Figure 9. The northern most exposure at Hells Half Acre volcanic field belongs to Twentymile Rock. Twentymile rock appears to be a lava tube based on its linear shape.

Figure 10. Hells Half Acre lava field source (arrow) is positioned on the horizon to the south. The lava field and its structures are displayed in the foreground.

Twentymile Rock is 1/8 mi off US 20 to left. This northernmost exposure of the Hells Half Acre field is now a National Historic Site and was a landmark to early settlers as they crossed the eastern Snake River Plain (**Figure 9**).

The highest point on the skyline to the left (south) is the vent area for the Hells Half Acre lava field (**Figure 10**). A small lava cone appears at the 3:00 position from this vantage point.

Road to the right leads to the crest of Microwave Butte. Microwave Butte is at the intersection of the axis of the ESRP and the Lava Ridge-Hells Half Acre volcanic rift zone. At the positions of 10:00 to 11:00 are East Butte, Middle Butte, and Big Southern Butte rhyolite domes that project above the surrounding basalt lava flows. Road to left leads to television transmitters atop East Butte (**Figure 11**).

Figure 11. View to the south from US 20 past Hells Half Acre are a set of volcanic cones marking the boundary rift between the ESRP and Hells Half Acre rift zones.

This is the boundary of Idaho National Engineering Laboratory (INEL). INEL, formerly called the National Reactor Testing Station was established in 1949 so that the U.S. Atomic Energy Commission (AEC), later the Energy Research and Development Administration (ERDA) and now the Department of Energy (DOE), could build, operate, and test various types of nuclear reactors. More than 50 reactors have been built to date. INEL occupies 894 mi2 (2320 km2). Road to the right leads to Argonne National Laboratory reactor and related facilites ("ANL").

Small tree-covered hill on the horizon to right of East Butte (9:00 position) marks the top of an unnamed low rhyolite dome that is mostly covered by basalt lava flows. Basalt flows that dip southward cap the Middle Butte. The unnamed rhyolite dome may be difficult to observe from US 20 due to the low elevation of the road. A small unnamed lava cone appears at the 3:00 position at this location. Cedar Butte appears at the 10:30 position left of Big Southern Butte. Road to right leads to Auxiliary Reactor Area (**Figure 12**).

Figure 12. Cedar Butte is a low profile volcanic vent which lies to the left of Big Southern Butte as viewed from west bound US 20.

The highway begins to descend onto Rye Grass Flat just past the East Butte access point. A little further along US 20, a turn out is located on the south side of the road with sign markers. This stop is designed to review the origin and volcanic history of the ESRP as outlined in the Introduction. The local geology including East, Middle, and Big Southern Buttes are pointed out and related to the regional geologic framework. The East and Middle Buttes are the two steep-sided hills that rise as much as 350 m above the surrounding basalt lava covered terrane. An unnamed hill about 1 km southwest of East Butte represents the apex of another rhyolite dome. Rhyolite lava flows and breccias are exposed at East Butte and at the unnamed hill, but not at Middle Butte. The steep sides of Middle Butte are covered by thick accumulations of talus composed of blocks that have been dislodged from the approximately 75 m thick layer of basalt lava flows that cap the butte. However, magnetic and gravity data suggest that the core of Middle Butte consists of rock that is less dense and less magnetic than basalt, and is probably rhyolite.

These factors and the structural data described below all indicate that the three hills are the upper parts of rhyolite domes. An east-west cross section through Middle and East Buttes is based on mapping, a drill core, and interpretation of geophysical data (**Figure 13**).

Figure 13. The turn out on the south side of US 20 just past East Butte (just to the left of the photo) marks the presence of a rift zone marked by the Middle Butte, Cedar Butte (hidden behind the road sign), and Big Southern Butte. A pair of signs at the turn out explains the viewpoint.

The internal structure of East Butte is known from study of the orientation of the flow layering in its rhyolite lava flows. The flow banding generally defines inward dipping concentric layers similar to those of a short stubby carrot or the lower half of an elongated onion. Flow layering parallel to an original upper surface occurs in many rhyolite domes but only locally at East Butte. The orientation of flow layering suggests that East Butte is a protrusion of lava that was too viscous to flow, and that the magma rose as inclined concentric sheets.

At the 12:00 position, Cedar Butte is a unique volcanic complex on the ESRP. The vent area consists of a circular pyroclastic cone more than 100 m high and 1.2 km in diameter and an elongated arcuate vent 1 km long and 30 to 60 m high. The arcuate vent extends north-northwest from the base of the pyroclastic cone. Field and petrographic data show that the two vents were active at different times with the pyroclastic cone being younger in age. Lava flows at Cedar Butte have a range in SiO_2 content from 54 to 67 percent. Thus, they are compositionally similar to flows of the COM lava field. A K-Ar age on a flow from Cedar Butte is 400,000 yrs.

At the 1:30 position, Big Southern Butte is the most prominent landmark in the ESRP. It rises 760 m above the surrounding lava plain. The butte is oval in plan, about 6.5 km in longest dimension, and elongated in a northwest-southeast direction. Big Southern Butte is a volcanic dome complex that consists of two coalesced cumulo-domes. A 350 m thick section of basaltic lava flows is exposed on the north flank of the butte. The basalt section was uplifted and tilted by the rise of the volcanic dome. Three K-Ar ages and one fission track age show that Big Southern Butte is about 300,000 yrs old. Faults, grabens, Big Southern Butte, Cedar Butte, and the vent for the Cerro Grande lava field are structural and volcanic features that define the Arco volcanic rift zone.

An electrical sounding profile across the ESRP from Blackfoot to Arco suggests an upper crustal structure in this area that consists of an upper layer 1.5 to 5 km thick of basaltic rocks, and a lower layer of sedimentary and rhyolitic rocks of unknown thickness. A 3,160-m deep exploratory geothermal test well CINEL-1 was drilled at INEL in 1978. More than 2,400 m of rhyolite ash flow tuffs, rhyolite lava flows, and interbedded air fall ash material are present in the lower part of the well. Less than 1 km of basaltic lava flows and interbedded continental sediments overlie the rhyolite section. A conclusion was arrived at that the well penetrated caldera fill material. The geographic relations of the Arco volcanic rift zone, the axis of the ESRP, the Lava Ridge-Hells Half Acre volcanic rift zone, and other localities near this stop are referred to in the text.

Return to the highway and head west. The junction of U.S. Highways 20 and 26 occurs north of Atomic City. Continue west on combined Highways 20-26 toward Arco. Highway 26 to the left leads back to Interstate 15 at Blackfoot. The highway leading to Arco now crosses over alluvium of the Big Lost River.

Road to the left leads to EBR-1 (Experimental Breeder Reactor-I), the first reactor built at INEL. Construction of EBR-1 was begun in 1949 to test a plutonium-fueled reactor for electrical power generation. The reactor was decommissioned in 1964, and in 1966 the facility was designated a Registered National Historic Landmark open to the public. Borrow pit to the left is in alluvial gravels of Big Lost River. Big Lost River Rest Area is on the left (**Figure 14**). The highway leaves the Big Lost River flood plain west of rest area. Crater Butte appears at the 11:00 position at the rest area entrance (**Figure 15**).

Road to the left leads to the spectacular crater vent in Crater Butte. US 20-26 junctions with Idaho State Highway 33. Continue on Highway 20-26 towards Butte City. There is an old truck weight station turnout on the right side past the intersection, a good place to stop and observe the surrounding area. Six Mile Butte belongs to a broad shield volcano at the 9:00 position at the intersection of an unimproved road leading south from US 20-26. A white house is set off the interstate intersection as a marker for identifying the road.

Note the profile of cinder cones along the Great Rift in Craters of the Moon National Monument at the 10:00 position (the largest cone is Big Cinder Butte). Pioneer Mountains, north of the COMNM is at 11:00-1:00 positions (**Figure 16**).

Figure 14. The gravel pit in alluvium from the Big Lost River is visible from the highway. The entrance road into the pit is just off to the right of the image on the south side of the road.

Figure 15. View from the rest area on US 20-26 west toward Arco Idaho. Crater Butte is on the horizon to the left of the road.

Figure 16. View from an unnamed dirt road intersection with US 20-26 east of Butte City showing the volcanics on the horizon from this view point.

The Pioneer Mountains immediately north of COMNM consist chiefly of Eocene extrusive and intrusive rocks, and Paleozoic rocks which include Carboniferous sedimentary rocks. For the next several miles on the right, note large eastward verging folds in Carboniferous and Permian carbonate rocks in south-facing slopes of the Arco Hills (**Figure 17**).

The highway crosses over alluvial fans formed by streams emerging from Arco Hills on the right. Butte City is the next town before crossing railroad tracks into Arco. Turn left on Box Canyon Road west of Butte City Idaho. Travel to the end of the road about 1.1 miles south. The road turns into an unimproved dirt road. Follow the path to observe the Box Canyon graben shown in **Figure 18**.

Stop 3. Box Canyon Graben

Upstream (northwest) from this point, the river flows within and parallel to the Arco volcanic rift zone. Downstream it flows northeast then back southeast along the rift trend cutting into ESRP basalts forming Box Canyon. This stop is on the southwest margin of the Box Canyon graben, a linear NW-SE topographic depression that controls the Big Lost River and marks the northern extent of dike-induced structures in the Arco volcanic rift zone. From this point and southeastward to Big Southern Butte, Arco volcanic rift zone surface deformation features and vents are typical of those observed in volcanic rift zones in Iceland and Hawaii.

Figure 17. At the descent of US 20-26 into Butte City, the right side Arco Hills display tilted rocks representing folds and faults shown by the arrows. The unimproved road on the left provides access into these hills.

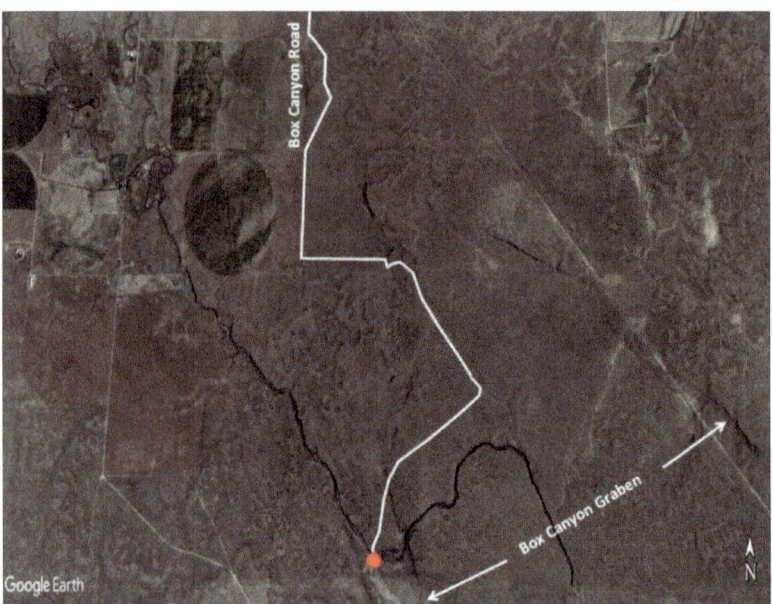

Figure 18. Box Canyon graben is a linear trending down dropped depression oriented to the northwest. The Big Lost River is structurally controlled by the Arco volcanic rift zone. Box Canyon follows the same general trend.

Fissures, fissure swarms, small normal faults, graben, eruptive fissures, and monogenetic shield volcanoes are abundant in this zone. From this point northwestward to the town of Arco, the origin of structures is ambiguous and transitional to the Lost River fault which is well developed north of Arco. The identification of late Quaternary fault scarps on the Lost River fault is clear and unequivocal from Arco to the north, but terraces along the Big Lost River and Holocene incision and deposition along a small drainage that joins the Big Lost River just south of Arco obscure any possible fault scarps and make identification of the fault uncertain. Reflection seismic surveys show displacements attributable to the Lost River fault for a distance of about 5 or 6 km south of Arco, but none in the area of the Box Canyon graben. Because of the uncertainty in fault identification, and the distance of possible epicenters of earthquakes to INEL facilities is sensitive to the position of the southern end of the Lost River fault. Several scenarios for the southern termination are used in seismic hazards assessment.

The Box Canyon graben probably represents the cumulative effects of several dike intrusion events into this part of the volcanic rift zone. The tilting of slabs of basalt and the fan-like separation of columnar joint blocks typical of deformation due to dike intrusion in volcanic rift zones worldwide is visible here in the area where the river cuts into the northeast limb of the graben. This ends the field trip into the Box Canyon graben.

Head back to US 20 on Box Canyon Road and turn left towards Arco, Idaho. In Arco, US 20-26 junctions with US 93. Turn left on the main street of Arco which is US 26-93, traveling southwest. The highway crosses the Big Lost River just past the intersection. The highway rises up onto a higher subtle alluvial terrace along Big Lost River, barely noticeable.

At the Arco Airport, the vent area for the Lost River Butte flows is at the 2:30 position. The flow margin is north of the highway and continues for about 1 mi (1.6 km) past the airport. The Lost River Butte flow rests on gravel-size alluvium of the Big Lost River. Just past the airport there is a turn out on the right which provides a view point on the right side of the highway (**Figure 19**).

Figure 19. The ridge past the turn out belongs to the Big Lost River Butte flows resting on Big Lost River gravels. The low rounded flow on the left side of the photo belongs to the Sunset flow of the Craters of the Moon.

The low rounded flow front on the left is that of the Sunset flow of the COM lava field. The 12,000-yr-old Sunset pahoehoe flows traveled more than 17.5 km from a source vent at Sunset Cone in COMNM to this site. The Sunset flow margin is sub-parallel to the highway all the way to Stop 4.

A tongue of Sunset flow crosses the highway. A second tongue of Sunset flow crosses the highway near the first tongue. Outliers of Ordovician-Kinnikinic Quartzite is viewed at top of the hill at the 3:00 position. The highway rises onto the Lava Creek flow of the COM lava field. The highway then rises onto the overlying Sunset flow where it is in contact with the Lava Creek flow (**Figure 20**).

STOP 4. Craters of the Moon Lava Field

Turn off onto the wide shoulder on the right. Walk east back to the contact of the Sunset and Lava Creek flows. At this stop, the first of four stops in the COM lava field, the two oldest flows at the north end of the field and their source vents are present. Flows about 15,000 yrs old lie at the nearly inaccessible south end of the COM lava field. The Lava Creek flow was erupted about 12,600 yrs ago from the southernmost Lava Creek vent along the Great Rift within the Pioneer Mountains. The flow cascaded about 500 m in about one mile to the level of the Snake River Plain at the site of the old town of Martin. It then traveled as an aa flow until it reached a point about 19 km from its source.

Figure 20. The contact between the Sunset and Lava Creek flow can be separated based on the color tones on the landscape. The darker tone represents the Sunset lava flow and the lighter tones surrounding it represent the Lava Creek flow. The view is facing back east after passing over the contact.

The Sunset flow is approximately 12,010 yrs old. It erupted from Sunset Cone at the north end of the Craters of the Moon National Monument. The Sunset flow consists of pahoehoe and its surface is hummocky with pressure ridges and plateaus (look south of highway), collapse depressions, ropes, and squeeze-ups. The Sunset flow is progressively more ash covered toward COMNM because dominant westerly winds blew tephra onto it during succeeding eruptions from Sunset Cone. Walk on the Sunset flow about 50 ft in from highway to observe the surface of this 12,000-yr-old pahoehoe flow. Blue glassy crust that is common on younger pahoehoe flows of the COM lava field is present here, but most has been obliterated by weathering, vegetation, and sediment cover. Most of the flow is obscured beneath vegetative cover presently.

The Lava Creek alkali basalt flow is hypo-crystalline and ranges from micro-porphyritic to porphyritic. Phenocrysts are mainly olivine and plagioclase. The latter are common near the south vent. The Sunset flow is hypo-crystalline and micro-porphyritic with micro-phenocrysts of olivine and plagioclase. Both the Sunset and Lava Creek flows represent the "primitive" eruptive products that are common in the early history of the COM lava field.

Figure 21. The Serrate flow exposes a jagged flow similar to an aa flow. The jagged lumps rising above the flow as cinder cone monoliths transported eastward, similar to pressure ridges formed within lava flows.

At Mile Post 232, junction with the road to Blizzard Mountain to the right. A flow with a serrated profile on the left is the Serrate flow. The jagged profile is due to cinder cone monoliths that the Serrate flow transported to the east when North Crater (in COMNM) was partially destroyed (to be discussed at STOP 5) (**Figure 21**). The boundary of Craters of the Moon National Monument is crossed at this point. On both right and left sides of the road are downwind cinder accumulations from Sunset Cone.

Stop 5. Craters of the Moon National Monument

At the entrance to Craters of the Moon National Monument, the Visitors Center is on the left. Stop for a snack and rest stop at the Visitors Center then follow the loop road into the Monument (**Figure 22**). At the entrance to the campground, cinder-covered latite block flow in the campground is one of the most felsic flows of the COM lava field, and it is considered to be part of the undated Highway flow (to be discussed later on) (**Figure 23**). A tongue of blue-crusted North Crater flow is visible on the left. This flow is one of several 2,100-yr-old basalt-hawaiite pahoehoe flows common along the loop road. We will visit the source of this flow in North Crater in the monument.

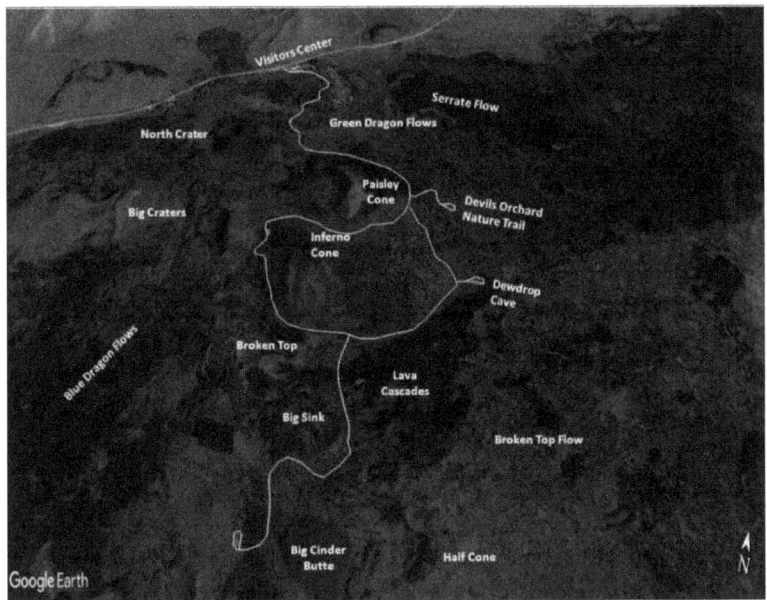

Figure 22. Aerial view of Craters of the Moon National Monument showing key features within the park including the Loop Road and side roads to sites. Image modeled after the National Parks site map.

Figure 23. Cinder covered (red brown) latite blocky flow (grayish brown) is visible in the center of the campground.

A turnout on the right is for the North Crater flow trail. Continue on Loop Road which crosses North Crater flow. Turnout on right is at the east flank of North Crater (**Figure 24**).

Figure 24. North Crater flow displaying pressure ridges and lava layering.

This is one of the trailheads to the North Crater-Spatter Cones trail. Continue on Loop Road. Big Craters flow on the right mantles the western flank of Paisley Cone (**Figure 25**).

Fissure vents for this pahoehoe and aa basalt-hawaiite flow are evident at the campground. For the next 1.3 mi, Loop Road follows the north, east, and south flanks of Paisley Cone.

Devils Orchard flow is on the left. This heavily cinder-covered latite block flow was thought by early workers to be quite old. It is correlated with the Highway and Serrate flows of eruptive period A. The Serrate, Devils Orchard, and Highway flows are considered to be quite young on the basis of stratigraphic and paleomagnetic data which will be discussed in more detail at a later stop.

Entrance to Devils Orchard Nature Trail is on the left. Continue on Loop Road. At the intersection with one-way Loop Drive, bear right. A tongue of the Big Crater flow is on the left in the valley between Paisley Cone on right and Inferno Cone on left. The vents for this flow will be seen at a later stop. Inferno Cone is at the 9:00 position; Big Craters cinder cone complex is at the 12:00 position (**Figure 26**);

North Crater is at the 2:00 position. The turnout leads to Inferno Cone parking lot. Turn right on the road leading to Spatter Cones parking lot.

Figure 25. Big Craters lava flow consists of pahoehoe and aa types. Vegetative coverage masks the lava structure but on the left side, pahoehoe structure appears as convex shaping towards the left.

Figure 26. From this view point, the profile of Big Craters is visible on the horizon. The dark brown structures in front of the left crater are fissure fountain spatter structures.

Park in the Spatter Cones parking lot. A 3.6 km walk to the Visitors Center will constitute this stop. The trail recently has been shortened considerably due to construction of the campground area. The hike is designed to view eruptive features and vent areas along a part of the Great Rift that was active about 2,100 yrs ago (**Figure 27**). A segment of the Great Rift 10 km long that extends from North Crater southeast to the Watchman cinder cone was active at various times during the latest eruptive period (A) of the COM lava field. The hike begins at the Spatter Cones, traverses the west side of the Big Craters cinder cone complex, crosses into North Crater, extends along the North Crater flow, and ends at the highway fault near the Visitors Center. The vents to be seen on the walking tour produced about 3.5 km3 of lava that now cover about 20 percent of the COM lava field. At about the same time as this activity, a rift segment of comparable length was active also and formed the Kings Bowl and Wapi lava fields to the south of the COM lava field. The Kings Bowl-Wapi area will be visited during a later part of Field Trip Exursion 2.

The Spatter Cones formed in the waning activity along a short (1 km long) eruptive fissure that extends southeast of the south end of Big Craters cinder cone. Most of the lava that forms the extensive Blue Dragon flow was erupted from the Great Rift in the Spatter Cones area. After viewing one spatter cone, take the trail to the west that ascends to south end of the Big Craters cinder cone complex (**Figure 28**).

Figure 27. Stop 3 trail starts at the Spatter cones parking lot and ends at the campground (dashed line). The length of the trail was cut short by construction of the campground.

Figure 28. Spatter cones erupt from fissures traversing south of the Big Craters. Source: Photographed by the author back in the early 1990's.

Big Craters is a cinder cone complex that contains at least nine nested cones. On the southwest rim, agglutinated spatter mantles the inner wall of the south and southeast parts of the complex (**Figure 29**). The mantle drapes over the rim of the complex and covers the outer wall (visible from a later stop). About 100 m north along the trail, remnants of a lava lake lie along the north wall of the inner crater (**Figure 30**). Just south of the lava lake remnant, the crest of a small cinder cone has a red streak aligned parallel to the eruptive fissure. Late-stage corrosive steam from the fissure oxidized the black cinders (**Figure 31**).

Lava issued from several satellite vents at the base of Big Craters cinder cone complex along its western (left) flank and travelled to the southwest. Additional nested craters are viewed in the Big Craters complex along the trail. As the trail descends the west slope of Big Craters, it passes near small craters on the west flank of the complex. Where the trail flattens out, it passes near a few small eruptive fissures to the left of the trail. The trail crosses eruptive fissures, source vents for the Big Craters flows in the area between the Big Craters cinder cone complex and the southwest flank of North Crater. Big Craters flows travelled both east and northwest from this area. This lava has an olive-green to greenish-brown crust which is useful in distinguishing Big Craters flows in areas where they abut younger and older flows.

Figure 29. The southernmost Big Crater is the deepest and largest crater of the trio. Agglutinated spatter mantles the inner wall at the south (center background) and southeast ends (right side) of the crater walls.

Figure 30. The middle vent of Big Craters contains a lava flow which is hidden from this view point.

Figure 31. The northern Big Butte Crater has reddish markers on the far left rim and in the left and right center walls and the right rim which marks the position of the fissure eruptive zone. The background hills belong to the North Crater.

The trail continues north to the west rim of North Crater cinder cone which has had a complex history. Clearly, North Crater cinder cone was larger than it is at present. Much of the northwest, north, and northeast flanks of North Crater was broken by slumping during a relatively explosive eruption of the Highway-Devils Orchard- Serrate group of flows, and crater-wall remnants were rafted away on flows from vents in and near North Crater (**Figure 32**).

Some of these crater wall remnants now appear as monoliths in the Serrate and Devils Orchard flows and as kipukas surrounded by the North Crater flow. From the crest of the trail on the west side of North Crater, slumped blocks and slump scarps can be observed on the northwest and northeast flanks of North Crater cinder cone. The trail descends into the vent area at North Crater. An entrance through a skylight into a lava tube is along the trail near the contact of the North Crater pahoehoe flow and the southwest part of the inner wall of North Crater. Farther east along the trail near the vent for the North Crater flow is a large block of agglutinated cinders that contains a granulitic xenolith. Such xenoliths, as well as xenoliths of pumiceous glass are common in cinders in the walls of North Crater. Leave the trail and walk north-northwest across the surface of the North Crater pahoehoe flow toward the highway. Ropes, hummocky surfaces, small monoliths, and the blue crust color of the North Crater flow are well displayed.

Figure 32. The North Crater lava flow is visible from the floor of North Crater. The large crack in the flow suggests a lava tube may be present beneath the surface.

Contacts between green-crusted Big Craters flows and blue-crusted North Crater flows can be observed near the steep scarp (called the "highway fault") parallel to and several hundred yards south of the highway. This fault possibly formed during collapse of the north flank of North Crater and the related eruption of the Highway-Devils Orchard-Serrate group of flows. Climb the scarp and walk to the Visitors Center for lunch.

Depart Visitors Center and then travel 2.7 mi along Loop Road to the parking lot at Inferno Cone.

The Inferno Cone parking lot is on the left. Climb the path up side of Inferno Cone. At top of cone, facing north, geographic features to be noted in a clockwise direction are: Big Craters cinder cone complex; North Crater cinder cone; Sunset Cone directly behind the Visitors Center with Pioneer Mountains in far background, and Paisley Cone. On the near side of the Sunset flow is the Serrate flow that extends to the east (right) as far as Round Knoll (off the image to the right). Round Knoll is a grass-covered kipuka of older Snake River Plain flows and cinders. At the "Snake River Plateau" display sign on the east rim of the summit of Inferno Cone, note Big Southern Butte and the surrounding terrane (**Figure 33**).

Figure 33. View looking north from Inferno Cone.

To the left of Big Southern Butte are East (left) and Middle (right) Buttes that appear to be one butte. Low shield volcanoes are clearly visible to the left and right of Big Southern Butte. The eastern part of the vast Blue Dragon flow is in the foreground. At the "Great Rift" display sign on the south rim of the summit of Inferno Cone facing southwest toward Big Cinder Butte note the following: the two easternmost cinder cones in this direction are Half Cone in the foreground and Crescent Butte with its distinctive crescent shape in the background. The dark saddle shaped cone is Blacktop Butte, the most southerly cone along the northern part of the Great Rift, 12 mi in the distance. Many of the more than 25 cinder cones in the COM lava field can be seen from this vantage point, but they are too numerous and closely spaced to differentiate them (**Figure 34**).

From the southwest rim of the crest of Inferno cone (no sign), one may observe the "plumbing system" that was responsible for the eruption of the vast eastern lobe of the Blue Dragon flow, the largest of all the 2,100-yr-old flows. The plumbing system consists of from the northwest to southeast: 1) an eruptive fissure located in the southern part of Big Craters cinder cone complex and beneath the area of the spatter cones; 2) pit craters such as Crystal Pit that overlie the southern end of the eruptive fissures; 3) perched lava ponds such as Big Sink waterhole that are located on the upper part of a lava tube system that extends east and south of the eruptive fissure; and, 4) a lava tube system that contains numerous skylight entrances into tubes (cave area along the Caves trail). Farther east are rootless vents where lava moving through the tube system was extruded through openings in tube ceilings.

Figure 34. The Great Rift Zone viewed from Loop road at the south side of Inferno Cone.

Also visible directly south beyond Big Sink waterhole and the Lava Cascades, is Broken Top cinder cone and the area to be visited at a later stop. Two 2,100-yr-old fissures slice across the northeast and southwest sides of Broken Top cinder cone. Source vents for the youngest flow in the COM lava field, the Broken Top flow are on the eruptive fissure on the east and northeast flanks of Broken Top.

Return to the parking lot. Turn left (one way) and proceed past the entrance to Spatter Cones parking lot. Bear left at the entrance to Spatter Cones parking lot. Big Cinder Butte is on horizon. Continue straight ahead. Blue Dragon slab pahoehoe lies to the right between the road and the spatter cones. The west flank of Inferno Cone is to the left. Turn right on the road to the parking lot for Tree Molds and Wilderness Trails.

Turnout on the right is a good place for viewing the Lava Cascades. Here, Blue Dragon lava flowed in a radial pattern from a lava lake in Big Sink waterhole, a perched lava pond (**Figure 35**).

Figure 35. Lava Cascades displays flow patterns where molten lava appears to have spilled over the Big Sinks area in the top center of the image.

The eastern rim of Big Sink waterhole lies several hundred feet west of the Lava Cascades parking area. On the left is a slab of Blue Dragon flow that mantles the north side of Broken Top. Either the flow was considerably inflated as it passed this site or the lava "sloshed" up onto the side of the cone as it changed direction from south to east. There are several other "slosh slabs" present in the Monument some of which do not seem to be explained by either of the two above mechanisms. Loop Road crosses the Blue Dragon slab pahoehoe (**Figures 36 & 37**).

A short walk (about 1.5 km) will constitute this stop. Park in Tree Molds and Wilderness Trails parking lot. Walk back on the road east to Buffalo Caves trailhead on the southwest side of Broken Top near the 2,100-yr-old fissure. Turn right (southwest) onto trail. Do not take the trail to left which ascends Broken Top. The trail to Buffalo Caves drops down into a fissure on the west side of Broken Top which has been filled by a tongue of Blue Dragon lava (**Figure 38**). Follow cairns on trail to southeast and walk on the surface of the Blue Dragon flow. To the left, the southwest-facing wall of the fissure has been mantled with spatter and bombs erupted from the fissure. Many faults that trend parallel to the Great Rift cut the west side of Broken Top (**Figure 39**). Walk along the trail to the contact of the Broken Top flow with Blue Dragon flow. Pahoehoe toes of Broken Top flow rests on top of Blue Dragon flow (**Figures 40 & 41**).

Figure 36. An exposure of "sloshed" Blue Dragon flow on the north side of Broken Top.

Figure 37. Pahoehoe flows crossing Loop Road on the west side of Broken Top.

Figure 38. Broken Top fault on the west side of the hill is represented by the hill slope scarp on the left. The canyon in the left center represents a fissure filled with Blue Dragon lava (white arrow).

The Broken Top flow, though not dated numerically is stratigraphically younger than the 2,076-yr-old Blue Dragon flow and therefore the youngest flow in the COM lava field. The Broken Top flow is mainly a lava lake in this area. Large slabs of the crust of the lake occur on the right side of the trail. The Broken Top flow is also blue crusted especially in squeeze ups that rise through cracks in the crust of the lava lake. The squeeze-ups appear to be from a younger flow.

The surface of the Broken Top flow has a relatively light color due in part to the weathering of its highly vesicular crust and to relatively greater amounts of vegetative cover than nearby flows. Continue on the trail (follow cairns) to Buffalo Cave. Buffalo Cave in the Broken Top flow shows several interesting features: lava stalactites, curbing (showing successive flow levels on the cave walls), ropes (showing flow direction), and floors formed by incomplete crusts of several levels of the lava stream.

What was the direction of movement of lava in the tube system? Where was the source vent(s)? Follow the southwest flank of Broken Top cinder cone east to a cinder path (the old Wilderness Trail). Turn left and take the path up southeast side of Broken Top. The Broken Top flows came from vents to the right of this path. At a very small borrow pit on the right side of the trail (on the north flank of Broken Top), turn left off the path and climb to the top of the cinder cone (**Figure 42**).

Figure 39. Fissure filled flow along Buffalo Caves trail on the south side of Broken Top. Flow belongs to the Blue Dragon lava.

Figure 40. In this example, Blue Dragon flow rests on Broken Top cinder (dark brown).

Figure 41. The Broken Top fault extends into the Blue Dragon flow along the trail in the southwest corner of Broken Top. The Blue Dragon flow is off set, easily observed from an aerial photograph. The large fracture to the left of the fault line probably separated during fault movement.

Figure 42. Vents where the Broken Top lava flow originated from at the southern end of Broken Top occur at the top of the ridge.

Spread out and walk amidst the sage and bitterbrush when making this ascent so as not to form foot paths on the side of the cone. The top of the cone is strewn with cinders and large (1 m) bombs erupted from the fissure on the west flank of Broken Top. At this point, a last overview of the Monument and a review of observations, data, and concepts are offered. Return to the northwest (toward Big Craters) to the cinder path and descend to the roadway and walk to the parking lot.

Leave Tree Molds-Wilderness parking lot. Turn right on Loop Road. Road to the Caves parking lot is on right. Continue straight on Loop Road. Devil's Orchard block flow is on the right. At the intersection turn right. The entrance to Devil's Orchard parking lot is on the right. Continue straight ahead. At the Visitors Center park for quick rest stop.

Leave COMNM, turning right (east) on Highway 20-26. Pass through Arco, turn right on Highway 20-26 in Arco and return to the intersection of U.S. Highways 20 and 26. At the intersection of U.S. Highways 20 and 26, turn right on Highway 26 toward Blackfoot. The highway crosses Rye Grass Flat, a grass-covered kipuka between Snake River Plain lava flows. Distal flow fronts of the Cerro Grande lava field is on the right. The Cerro Grande lava field is about 10,000 yrs old and was fed from a vent at the base of the southeast flank of Cedar Butte, the prominent butte immediately southeast of Big Southern Butte.

Stop 6. Cerra Grande Volcanic Field

At the Bingham County line on the right is the town of Atomic City. At Atomic City turn right onto either Twin Buttes Road or Main Road into the city. At Taber Road turn left (south) to the intersection of W 1600 N Road. Drive 1.3 mi and turn west onto an improved unpaved road. Continue 4.1 mi to the railroad crossing near the margin of the Cerro Grande lava flow (**Figure 43**).

Park on the west side of the tracks. Observe the flow margin of Cerro Grande lava and the effects of inflationary emplacement of pahoehoe lavas typical of ESRP basalts. The flow margin comprises two parts: a thin outflow "sheet" pahoehoe and an overlying thicker flow made of large inflationary pahoehoe lobes. Vesicles and mineral textures reflect variable pressures inside inflating and deflating lava lobes. Layers of vesicles occur in cycles often as 1-2 cm layers of intensely vesiculated lava separated by 10-20 cm of more massive lava representing repeated pressure increase during lobe inflation followed by decompression and volatile exsolution.

Figure 43. Southwest of Atomic City, the Cerro Grande volcanic field is present. Big Southern Butte is visible to the right as viewed from the road and smaller volcanic vents are present on the left.

Thickness of the crust increased with each cycle. The vesicle layers are curved and concentric with lobe surfaces rather than horizontally oriented due to gravitational rise of bubbles such as those noted for vesicle layers within Columbia River basalt flows. This suggests that molten lava within each lobe experienced cyclic velocity changes as magma repeatedly backed up during marginal cooling then broke out when internal pressure rose above the tensile strength of the crust. Similar processes are noted for sheet flows in Hawaii.

Climb up the 5-8 m lobe on the northeast side of the tracks where deflationary sagging has resulted in deep tension gashes along the periphery. From this point, one can observe cyclic vesicle layers in the fracture walls related to repeated decompression of the lobe interior as the crust thickened downward. The flow surfaces on the top and flank of the lobe are ropy pahoehoe typical of the thinner sheet flow on the lower level that were originally close to horizontal. This observation attests to the significant amount of inflation necessary to produce several meters of uplift due to internal magmatic pressure or gas pressure caused by volatile exsolution (**Figure 44**).

Figure 44. View facing northwest from the northeast position of the railroad tracks and unimproved dirt road of the Cerro Grande lava field.

Continue traveling west along the unimproved dirt road from the Cerro Grande lava field. The road curves to the southwest. A map is provided in **Figure 45**. In the southeast direction from about the northern third of the dirt road, Cedar Butte and its lava flows are visible.

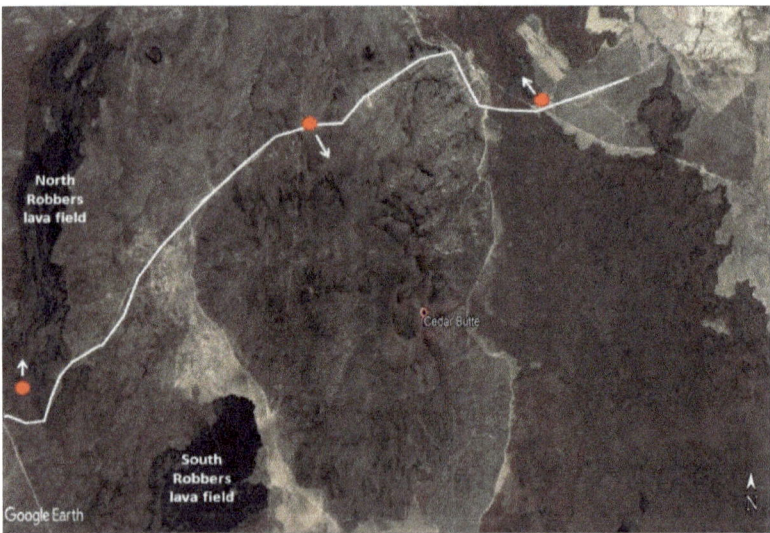

Figure 45. Aerial view showing route from Cerro Grande lava field to the view point of Cedar Butte volcanic center. The lower left red dot indicates the last stop in this area at the gap between North and South Robbers lava fields. White arrows indicate viewing direction.

Stop 7. Cedar Butte

Drive west from the railroad crossing on the improved unpaved road for 2.8 mi. Turn south on an unimproved dirt road and continue 1.5 mi to where the road splits. Take the left (east) fork and proceed 0.7 mi to the stop point located near the center of Cedar Butte volcano (**Figure 46**). Park on the dirt road near a breach in the northern side of a large tephra cone. The purpose of this stop is to discuss possible genetic relationships between evolved mafic-intermediate volcanic centers and high-silica rhyolites of the ESRP. Walk north a few tens of meters to the prominent bladelike exposures of mafic dikes. Here, you will see three en echelon dike segments. They are a part of a nearly continuous curvilinear system of vents and intrusions extending for 2 km, and circumscribing a 150 degree arc. The arc has a radius of curvature of about 0.8 km, and may have originated by incipient formation of ring fractures above a shallow magma chamber.

These dikes exhibit spectacular features of bimodal magma interaction. The dikes strike N8E and dip vertically. The overall dike morphology is wedge-shaped, tapering towards the top suggesting that the original magma had not propagated to the surface at this location. Contacts with the host tephra are bulbous on a scale of 10's of cm to a meter.

Figure 46. View facing Cedar Butte from the middle red dot in Figure 45.

Contacts with the host-rock tephra have produced a thin zone where the glassy tephra pyroclasts are fused into dense obsidian.

Both southerly dike segments have a layered structure consisting of a core of felsic composition and a rind of more mafic composition. The interior of the southerly dike is rhyolitic in composition. In the middle segment, the core consists of trachydacite. In both cases, the rind consists of trachyandesite. Contacts between the two lithologies varies from sharp to gradational over a few centimeters. In many places there are wispy blebs of one lithology within the other. Both features indicate the original magmas were at least partially molten when they were brought into contact. The dike lithologies overlap in composition with the compositions of pyroclasts in the tephra cone to the south suggesting that the dike is the northern continuation of vents which fed the tephra cone forming eruption.

Stop 8. North Robbers Lava Field

Return to the improved unpaved road, continue west 4.2 mi and turn right onto a service road. Drive as far as possible, approximately 0.3 mi to the last stop in this area which is located at the vent for the North Robbers lava field. This is one of the three smallest lava fields (along with South Robbers and Kings Bowl) in the late Pleistocene and Holocene episode of ESRP volcanism.

The fissure zone for North Robbers is 2.9 km separated into a 1-km-long northern segment which has both eruptive and non-eruptive components, and a 1.2 km long southern eruptive segment. The non-eruptive part of the northern segment extends northwest into the Big Southern Butte rhyolite dome. The southern segment is offset from the northern segment and is the primary eruptive fissure system for the North Robbers lava.

Hike up to the summit region which is constructed by a spatter rampart at the vent (**Figure 47**). This is a good vantage point to observe fairly uncomplicated eruptive features associated with ESRP basalts. On the north side of the cone, the lava channel flow extends north-northeast from the vent area. Walk around to the southeast side to observe the eruptive fissure system oriented northwest-southeast along the Arco volcanic rift zone from which a relatively short lobe of fissure fed lava that flowed northeast.

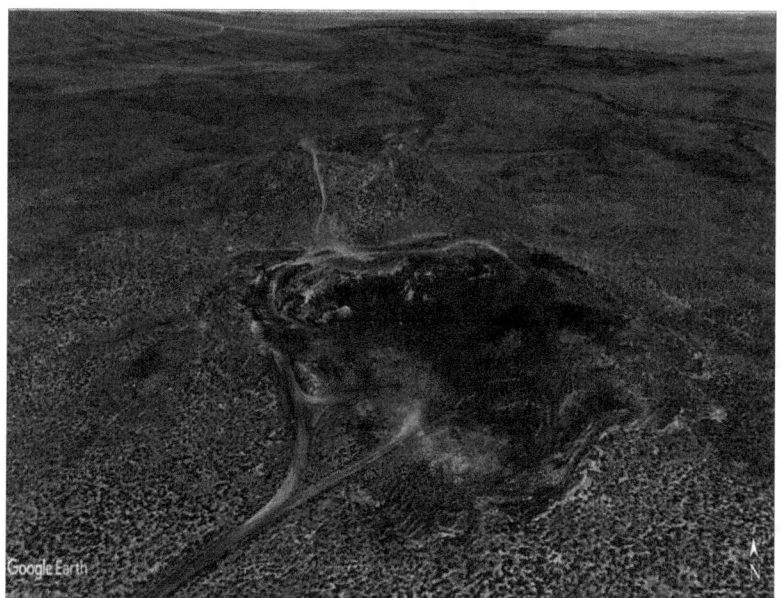

Figure 47. View of the North Robbers lava field. The North Robbers vent is in the upper center with a path leading to the summit. The foreground red area was exposed from off road vehicular traffic. The area appears to be an eroded volcanic cone associated with the Big Southern Butte. Volatile exsolution may be the cause of the red coloration.

This is the last stop in this area. Trace the dirt road back to Atomic City and return to US 26, turning right (southeast direction).

Table Legs Butte is on the right (**Figure 48**). This shield volcano has a high crowned vent due to lava lake activity. Middle and East Buttes are on the left. Low hills on the right represent rootless vents on a tube system in a flow that traveled east from Table Legs Butte (**Figure 49**). Borrow pit on the right shows accumulation of at least 1 m of loess on the Snake River Plain flow. Kipuka filled with loess is on the left.

Figure 48. Table Legs Butte is off in the distance on the right side of US 26 south of the intersection with US 20.

Figure 49. Lava tubes formed from Table Legs Butte are exposed along the right side of US 26 just south of the butte.

Ascend Taber Butte, a low shield volcano whose crest is at the 11:30 position. A low shield vent area of Taber Butte appears at the 9:00 position. Portneuf Range appears at the 1:00 position.

The highway crosses distal tongues of the Hells Half Acre lava field. Pressure ridges and pressure plateaus are common morphologic features of this 4,000 to 6,000 yr old pahoehoe flow. The Hells Half Acre lava field covers an area of about 430 km2, and contains about 3 km3 of lava. The lava field probably formed during a single eruptive event that lasted perhaps several months or a few years. The flow units that make up the lava field all flowed from a central vent complex about 1,000 m long and 300 m wide. The vent complex consists of an elongated crater that contains remnants of lava lakes. A few small cinder cones, aligned parallel to an eruptive fissure system lie in the crater and beyond its edges. An extensive lava tube system reaches south and east of the vent area (**Figure 50**).

Figure 50. Pressure ridge from Hells Half Acre flow is exposed at the 818 address of US 26 in Blackfoot.

Lava flows of the Hells Half Acre lava field consist of olivine tholeiite. U.S. Highway 26 drops off Hells Half Acre flows.

The highway crosses Peoples Canal. Gravel-size alluvium originating from former Snake River courses are common along the sides of irrigation canals. This area was inundated by the 1976 flood caused by the failure of Teton Dam. The highway crosses Snake River.

At the intersection of U.S. 26 and 1-15, turn right on the entrance ramp to 1-15 heading south on 1-15 toward Pocatello. Blackfoot is on the left after turning on the interstate highway.

At the overpass on Willie Road, the left side is Blueberry Hill. This broad round hill (diameter of 9 km) probably overlies a rhyolite dome, but no rhyolite is exposed. Basalt flows dip outward on its north, west, and south flanks. The broad low shield volcano on the right is Ferry Butte, a landmark for early travelers on the Oregon Trail. This was one of the few safe places to ford the Snake River.

Fort Hall and Fort Hall Indian Reservation is to the right. Continue on 1-15. In the early 1800's, Fort Hall was a trading post in this part of Idaho. Pocatello Range is on the left, Michaud Flats on the right. Pocatello Range consists chiefly of a basement complex of Precambrian quartzite, sandstone, argillite, and meta-volcanic rocks. Several thrust sheets have been recognized. The Precambrian rocks are overlapped by Miocene rhyolite tuffs of the Starlight Formation. Michaud Flats is a fan-delta that consists of boulders, sand, and gravel deposited by the Bonneville floods upon its emergence onto the Snake River Plain. At the junction of I-15 and 1-86, this ends the first day of the field trip.

Field Excursion 2. The Southern Great Rift

Champion and others (1983) provided a field excursion into the Kings Bowl and Wapi Lava Field as part of the Great Rift discussion.

Geologic Overview

The volcanic features to be observed during this field trip excursion contrasts with the features observed during the field trip excursion 1. The Kings Bowl and Wapi lava fields are relatively simple products of single eruptive bursts. They are not compound lava fields that record multiple eruptions such as the COM lava field. However, there are also marked contrasts between the Kings Bowl and Wapi fields that reflect different styles of eruption for each field. The Kings Bowl lava field is small and accumulated adjacent to a 7-km-long eruptive fissure segment of the Great Rift. The Wapi lava field is larger and probably began initially as a fissure eruption, and with more prolonged activity progressed to a sustained eruption from a central vent complex. The compositions of both Kings Bowl and Wapi lava flows resemble the average olivine tholeiite of the Snake River Plain in contrast to the evolved COM lava flows from the northern part of the Great Rift.

Kings Bowl Lava Field

The Kings Bowl lava field consists of about 0.005 km3 of pahoehoe lava flows that cover an area of about 3.3 km2 along the southern part of the Great Rift 12 km southeast of the Craters of the Moon lava field. The lavas were erupted about 2,220 ± 100 yrs ago from a 7 km long central fissure of the Kings Bowl rift set. The set consists of a central eruptive fissure that trends N10°W flanked by two sub-parallel non-eruptive sets of cracks that are from 600 to as much as 1,100 m from the main fissure. Flanking cracks are older than the central fissure, and typically are less than 1 m wide. The central eruptive fissure consists of discontinuous linear en echelon cracks 2 to 3 m wide that locally are filled with breccia and feeder dikes.

The Kings Bowl lava field is characterized by thin (typically < 0.2 m) fissure fed pahoehoe lava flows, lava lakes, low natural levees, spatter ramparts, spatter cones, and explosion pits. Large spatter cones and spatter ramparts adjoin explosion pits at South Grotto and Creons Cave, and light-colored blankets of lapilli tephra spread eastward from the explosion pits. The largest explosion pit on the Kings Bowl rift is Kings Bowl, 85 m long 30 m across and 30 m deep. A lava lake surrounded Kings Bowl before the explosive eruption. Prominent basalt mounds believed to be the remnants of levees define the limits of the lava lake.

Blocks as large as 10 cm in diameter were hurled explosively westward as far as 245 m. The lapilli ash tephra resulting from the explosion was carried eastward by the prevailing winds. Tephra, about 1 mm in diameter occurs as far as 1.2 km east of Kings Bowl. The tephra blankets an area of 0.15 km2. As the larger blocks that resulted from the explosion fell on the west side of Kings Bowl, many broke through the crust of the lava lake and through squeeze-ups. At some localities, the impacting projectile can be found in place beneath the crust. Calculations show that the volume of the ejected material falls far short of that needed to refill the cavity of Kings Bowl indicating collapse in the vent area subsequent to the explosive vent.

Wapi Lava Field

Unlike the COM lava field, the Wapi lava field is a monogenetic volcano and typical of many of the low shield volcanoes that make up most of the present surface of the central and eastern Snake River Plain. Thus, study of the Wapi low shield is important in understanding the processes involved in the formation of this type of volcano. The Wapi field covers a 326-km2 area that is elongated north-south along the Great Rift. The margin of the field is smooth on the north and east sides where it ponded against the regional slope of the Snake River Plain. On the south and west sides, the margin is formed of long lingular flows where they filled drainages of small intermittent streams. The slope of the Wapi lava field over distances of 10-20 km is less than one degree. The flat slope is a consequence of 1) the very fluid pahoehoe flows; 2) the relatively high rates of lava effusion; and, 3) the original flat slopes over which the flows moved. The only area of the lava field that has a steeper slope is near the vent area of Pillar Butte where the slopes range from 5° to 7°.

The Wapi field is composed of numerous flow units of pahoehoe lava that are piled side by side and atop one another. This forms a type of lava flow described as "compound." Exposures in many kipukas along the south and west sides of the lava field suggest that the flows there are about 5-10 m thick. Thicknesses of the Wapi flows are 15-25 m except at Pillar Butte where the total thickness may be 100 m. Near the margins of the field, the flow units are larger and tend to have a greater local relief (as much as 10 m) and are characterized by large pressure plateaus, flow ridges, and "collapse depressions." The transition in the size of the flows from the periphery to the interior of the field is apparently a function of proximity to the vent area. Thus, closer to the vent many small pahoehoe flows have filled depressions in earlier flow units and generally leveled the local relief.

The numerical age of the Wapi lava field was unknown until recently. Wapi lava flows overlie and flowed into open cracks of the King's Bowl rift set which in turn produced the Kings Bowl lava field. The weighted average of three radiocarbon ages on charred sagebrush found under the Kings Bowl lavas is 2,220 ± 100 yrs B.P. A sample of charred root material was obtained from an excavation under the Wapi lavas at the east edge of Wapi Park. A radiocarbon age on this sample was 2,270 ± 50 yrs B.P. Thus, the Wapi and Kings Bowl lava fields formed simultaneously. The contemporaneity of the Wapi and Kings Bowl lava flows is supported by paleo-magnetic data.

Precise directions of remnant magnetization obtained from two separate outcrops (17 km apart) within the Wapi field agree closely with each other and with the direction of magnetization of lavas from the Kings Bowl field. In addition, the paleo-magnetic data suggest that the Wapi field took a short time to form, probably several months to a few years despite the large volume (approximately 6.5 km3) of basalt erupted.

Remnant magnetization directions of lava flows near Pillar Butte differ from those from lava flows of the Wapi lava field. The difference can be attributed to rotation associated with summit deflation near Pillar Butte at the close of the eruption, or to a younger age of lava flows near Pillar Butte than the age of the main part of the Wapi field. The latter possibility seems unlikely in that flows near Pillar Butte both underlie and overlie flows of the rest of the Wapi field.

Surface morphologies of the Wapi lava field are readily observable and are characteristic of low shield volcanoes of the Snake River Plain. Near Pillar Butte, thin flows were fed on the surface from the vent complex. Leveed-channel aa flows, shelly pahoehoe rootless flows, and pahoehoe flows composed of toes are common. Other lava flows of the Wapi field are pahoehoe in surface texture and were fed by tubes. Ubiquitous features are "pressure ridges" and "collapse depressions." Unfortunately, both features are incorrectly named with respect to mechanisms of formation. True pressure ridges in lava flows are analogous to pressure ridges of sea ice that form from wind pressure pushing up ridges transverse to the direction of the wind. The lava ridges of the Wapi lava field are parallel to the direction of lava flow. Thus, the term "flow ridge" is used here in contrast to "pressure ridge." Pressure ridges have been observed to form on the surface of a lava lake in Mauna Loa. The ridges on the McCartys lava field in New Mexico were recognized flow ridges attributed to collapse of the inflated crust over a lava flow.

Collapse depressions have been thought to form by collapse of the roofs of lava tubes. In a study of 100 collapse depressions, none were found to be related to an un-collapsed lava tube system. From their shapes and positions, it is apparent that the collapse depressions of the Wapi field are instead related to an internal flow system. The gentle slopes of the Wapi field make it seem unlikely that a hydrostatic head could exist to drain tube systems.

Observations of the pattern of flow preserved in surface textures around collapse depressions, the morphology of the depressions, and the striations on inward-facing scarps of depressions suggest that depressions form early in the cooling of the flow, and not from collapse. Rather, they can be thought of as localities where the lava crust never was inflated.

Bimodal Magmatism, Basaltic Volcanism, Tectonics, and Geomorphic Processes of the Eastern Snake River Plain.

Hughes and others (1997) assembled a field trip through bimodal volcanics from Pocatello through American Falls to the Kings Bowl and Wapi Lava Field along the Great Rift zone.

Geological Overview

Geology presented in this field guide covers a wide spectrum of internal and surficial processes of the eastern Snake River Plain, one of the largest components of the combined late Cenozoic igneous provinces of the western United States. Focus is on widespread Quaternary basaltic plains volcanism that produced coalescent shields and complex eruptive centers that yielded compositionally evolved magmas. The guide is constructed in several parts beginning with discussion sections that provide an overview of the geology followed by road directions with explanations for specific locations. The geology overview briefly summarizes the collective knowledge gained, and petrologic implications made over the past few decades. The field guide covers plains volcanism, lava flow emplacement, basaltic shield growth, phreato-magmatic eruptions, and complex and evolved eruptive centers. Locations and explanations are also provided for the hydrogeology, groundwater contamination, and environmental issues such as range fires and cataclysmic floods associated with the region.

Massacre Volcanic Complex west of Pocatello, the 1-86 freeway roughly parallels the Snake River along the southern physiographic boundary between Basin and Range and Snake River Plain geologic provinces. Late Tertiary and early Quaternary volcanic rocks partly fill the intermontane valleys of the Basin and Range and lap onto their adjacent mountain flanks.

Downstream from American Falls, the Snake River cuts through late Miocene and early Pliocene rhyolite deposits that are overlain by a complex assemblage of Neogene basaltic tuffs and lava flows. The Neogene units were redefined as the Massacre Volcanic Complex composed of the Eagle Rock, Indian Springs, and Massacre Rocks basaltic pyroclastic sub-complexes, and the basalt of Rockland Valley.

The pyroclastic complex crops out over a 128 km2 area about 15 km southwest of American Falls. The flows were deposited during explosive phreato-magmatic eruptions along the Snake River. The Massacre Volcanic Complex overlies a sequence of unconformable volcanic and volcani-clastic deposits including (youngest to oldest) the Little Creek Formation, the Walcott Tuff, and the Neeley Formation.

The trip begins at Pocatello Idaho at the junction of Interstate 15 and Interstate 86. Traveling south from Pocatello on Interstate 86 take the Register Road exit, turn right. At the intersection with Park Road, turn right again. Park Road leads into Massacre Rocks State Park. Approaching the exit ramp, a road cut exposes a portion of the Massacre Rocks basalt which is highly fractured (**Figure 51**).

The northbound lane also has a road cut belonging to the same flow. At least two sub-horizontal Quaternary basalt lava flows which flowed northward down Rockland Valley from Table Mountain shield volcano lap onto basaltic tuffs of the Massacre Rocks sub-complex that have a slight westward dip. Auto-clastic basal breccia and baked paleo-sol occur between the two lava flows. The underlying unit is a stratified lithic tuff with subtle gas escape features that are perpendicular to the dip. These elutriation pipes, assuming they were originally vertical, possibly indicate post-eruptive tectonic tilting or local down warping along the southern margin of the ESRP.

Continue on the freeway and take Exit 28, Register Road exit. Make a left and go over the interstate to the frontage road at the first left. Traveling northeast on the frontage road for about one-quarter mile, juniper-covered slopes about 0.5 km south of Massacre Rocks State Park appear. Hike up to one of the prominences about 100 m off the road to observe bedded tuff deposits and ballistic ejecta.

Figure 51. Stop 1A is a portion of a roadcut in the southbound lane of Interstate 86 before the Register Road exit. The north bound side of the interstate (behind the photograph view) also has the eastern portion of the road cut visible.

The deposit is on the flank of the Massacre Rocks unwelded tuff cone. Block sags are common and reflect close proximity (less than 1 km) to the vent. Scour surfaces and well-stratified planar bedding and dunes indicate high flow velocities during volcanic surges which are fairly common in ESRP phreato-magmatic deposits.

Return to the freeway continuing northeast. Take Exit 33 and re-enter the freeway headed southwest to the rest area near Massacre Rocks State Park. Walk southwest from the rest area parking lot along the paved path leading to the old Oregon Trail wagon ruts. The path forks and the paved part passes under the freeway to the Oregon Trail exhibit. Continue southwest on the unpaved path for about 100 m. Leave the path and climb down the ridge toward the river to observe the stratigraphic section. The sequence that can be observed at this location on the opposite side of the river consists of of Cedar Butte basalt, a Quaternary ESRP lava flow and ignimbrite of the Massacre Volcanic Complex is visible (**Figure 52**).

The Neely Formation consists of poorly sorted massive tan poorly indurated friable fine tuff with some calcite nodules. White fossiliferous marl is near base at the level of the Snake River. The Walcott Tuff overlies the Neeley Formation in sharp conformable contact, a 6.5 Ma ash-flow tuff. An upper obsidian welded tuff grades from weakly to densely welded at the top.

The central zone consists of densely welded vitro-phyre with spherulites grading to welded tuff at the base. The lower part is a bedded tuff, white to light gray, planar bedded, medium to fine vitric crystal tuff. The lower 1.5 m locally contains accretionary lapilli with the upper 0.3 m locally fused by the overlying welded tuff. An unconformity occurs with the overlying Little Creek Formation consisting of brown to gray massive basaltic tuff breccia, yellowish brown lapilli tuff, tan to yellow tuffaceous palagonitic sandstones and calcareous siltstones. The basal part is interpreted as being of colluival origin, but may possibly be fluvial depositional at the inception of the Massacre volcanism. On top, the Massacre Volcanic Complex lies within a deeply channeled unconformity above the Little Creek Formation. It is late Miocene to early Pliocene red brown to light gray basaltic tuff breccias, tuffs, dikes, plugs, and flows. It is unconformably overlain by Cedar Butte basalt in places. Some tuff breccias similar to the underlying Little Creek formation occur.

Figure 52. Stop 1C is the gullied exposure at the Massacre Rock rest area on Interstate 86. Exposures of basalt reveal Neeley Formation un-welded ignimbrite or reworked debris flow (?) possibly late Miocene.

American Falls to Kings Bowl and Wapi lava fields and return

Starting at Idaho 39 at the American Falls Dam, travel north to North Pleasant Valley Road. Take North Pleasant Valley Road to Winters Road and make a right on Winters Road. Travel north on Winters Road to Crystal Ice Cave Road. Make a left turn on Crystal Ice Cave Road, and travel northwest. Crystal Ice Cave Road is an unimproved road so a 4 wheel drive is recommended during the rainy season.

Several miles, there is a turn off on the right. Go straight ahead. This is Stop 1, King's Bowl (**Figure 53**). The Queens Bowl crater is to the right. This explosion crater is located on a north-trending rift system that is older but similar in appearance to the Kings Bowl rift set (**Figure 54**). Flows of the Kings Bowl lava field are to the right of King's Bowl.

STOP 1. Kings and Queens Bowls

Park at the Crystal Ice Cave parking area. At this stop, examine the Kings Bowl lava field and discuss its eruptive history and relationship to the Great Rift.

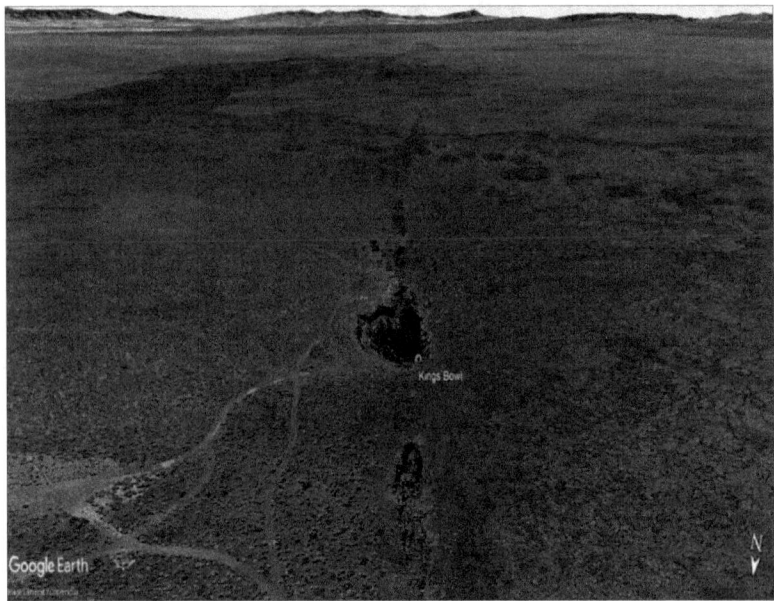

Figure 53. King's Bowl is a long fissure that produced the lava flow. The fissure is part of the Great Rift. The bowl is a collapse above the rift with smaller collapse zones along the rift's length.

Features to be observed while walking include the main eruptive fissure, spatter cones, spatter ramparts, lava lakes, lava levees, basalt mounds, the Kings Bowl explosion pit, and effects of the explosive event at Kings Bowl. Walk to top of the lookout directly west of Crystal Ice Cave parking area. The view to the west and southwest is of a former lava lake surface. North and west of the lava lake are several basalt mounds believed to be remnants of levees that contained lava lakes.

Proceed northward from the lookout along the main eruptive fissure to an area where age relationships of Kings Bowl lava flows and non-eruptive fissures of the Great Rift are apparent. Cracks cut the older flow but are overlapped by the younger flow (**Figure 55A**).

Proceed west across the main eruptive fissure to an area of levees that contained a lava lake. Lava tubes and lava channels that resulted in overflow of the levee system also can be observed here (**Figure 55B**).

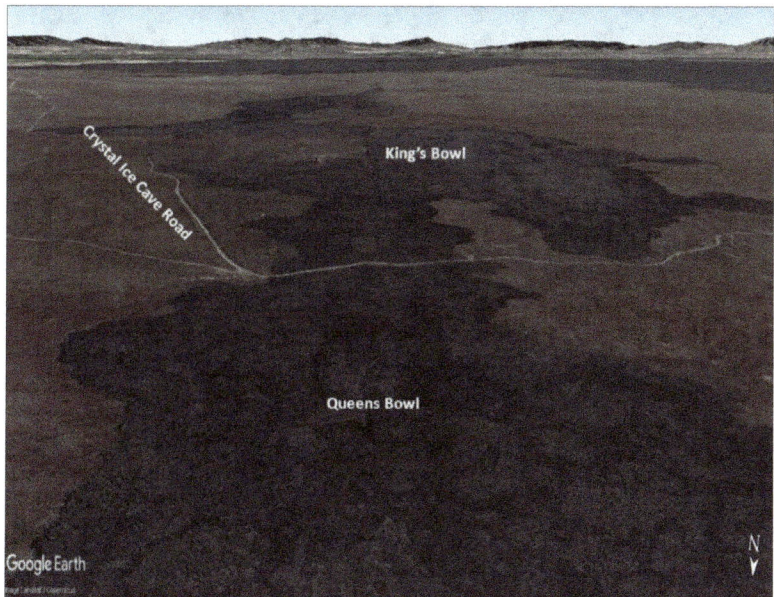

Figure 54. Queen's Bowl is north of King's Bowl, an older part of the Great Rift Zone.

Proceed southward along the main eruptive fissure to an area with squeeze-ups, ejecta blocks, basalt mounds, and surface of the lava lake. "Flow-out" features can be seen between the basalt mounds. The lava lake is flat at this locality and it is broken into large plates as a result of subsidence of the lake. Bulbous squeeze-ups resulting from lava oozing through cracks between plates on the lava lake surface are abundant. Lithic blocks thrown out by the explosive event at Kings Bowl litter the surface (**Figure 55C**). Impact craters in the lava lake surface from such blocks can be observed.

Proceed eastward across the main eruptive fissure to the locality at the north end of the Kings Bowl crater near the tourist wall where a feeder dike intrudes the main eruptive fissure (**Figure 55D**). Follow the trail southward around the east side of Kings Bowl crater to the crater entrance. Here, the surface east of the main eruptive fissure has been blanketed by fine tephra erupted during the phreatic explosions at Kings Bowl.

At the crater entrance, the upper flows display a swirling pattern as a flow variety characteristic of near-vent activity (**Figure 55E**). Farther down the trail, a soil horizon separates the young Kings Bowl lava flows from older massive flows in the crater wall (**Figure 55F**). Continue down the trail to the north end of the Kings Bowl crater where a feeder dike is exposed in the central eruptive fissure. Walk north into the fissure to observe the grooved walls resulting from phreatic eruptions. Retrace the trail out of the crater, turn north and return along the trail to the parking area.

Figure 55. Viewpoints around King's Bowl where features are described in the text.

Leave Crystal Ice Cave parking lot on Crystal Ice Cave Road traveling to the southeast. Parts of the next segment of the field trip require a high centered or 4-wheel drive vehicle, particularly in wet weather. Turn sharply right onto a dirt road. Proceed first west, then curve south. Turn right onto the dirt road and proceed nearly due west. In the near distance, spatter vents and near-vent flows of the southern part of the Kings Bowl lava field are present. Continue west past the road on the right (**Figure 56**).

STOP 2. Kings Bowl Fissures (Optional)

This area constitutes an optional stop (time permitting) to observe fissures of the Kings Bowl segment of the Great Rift. Extensional deformation produced fissures as much as 0.5 m across at this locality. These fissures belong to the eastern set of three parallel fissure sets. The eastern set of fissures is separated from fissures of the central set by about 0.7 km. Systems of nearly parallel fissures also characterize the area immediately southeast of the COM lava field.

Although the Kings Bowl fissure system trends N10W, the individual fissures trend N22W, in an en echelon manner. This trend suggests a component of left shear in this area of the Great rift (**Figure 56).**

Proceed west and cross the Great Rift and eruptive fissures that produced the Kings Bowl lava field. Spatter cones occur both north and south of the road at this point. Turn left and proceed south on a dirt road. The road traverses the length of Wapi Park, a vent complex that trends southwesterly, older than the nearby Wapi and Kings Bowl lava fields (**Figure 57**).

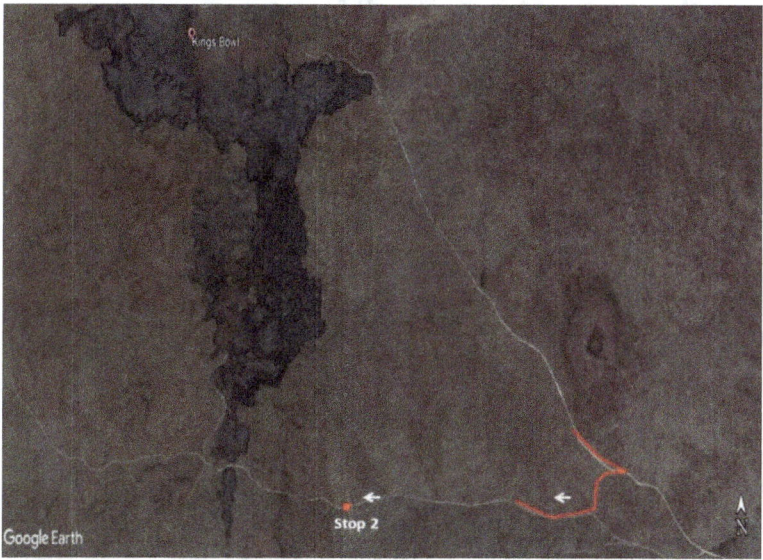

Figure 56. Route map from King's Bowl southeast on Crystal Ice Cave Road to turn off at the dirt unimproved road traveling west.

Turn right and proceed south-southwest between the edge of the Wapi lava field and the old vents of Wapi Park.

Proceed south-southwest past a left turn. A hand-dug tunnel here yielded charred sagebrush beneath lava flows of the Wapi field whose radiocarbon age is 2,270 ± 50 yrs B.P.

One of the paleo-magnetic sites for the Wapi lava field is located on the inflated pressure plateau to the west of the road. Data are the same for the Pleasant Valley arm of the Wapi field which projects from the eastern margin of the field, 20 km to the south.

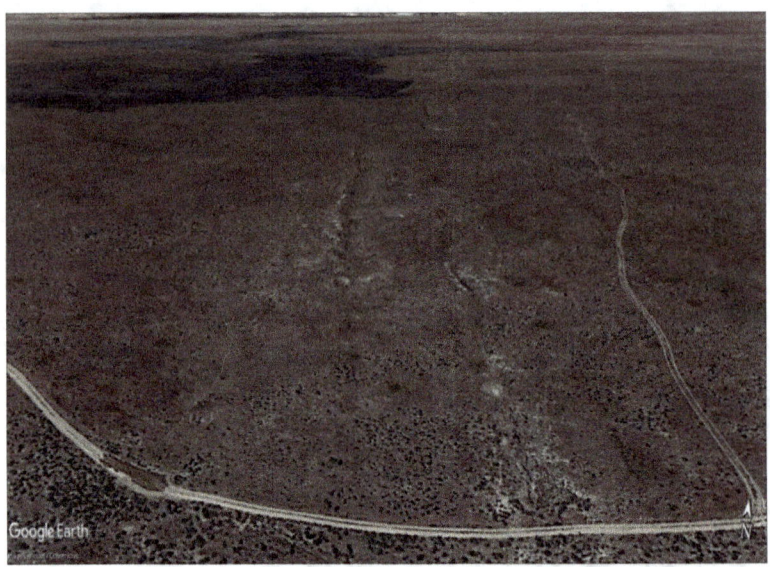

Figure 56. Great Rift fissure produced by extensional deformation appear as linear depressions on the landscape surface. Fissures were produced by left shear in this part of the rift.

Figure 57. Wapi Park vent complex located south of Kings Bowl.

STOP 3. Wapi Park Lava Field

Travel to the end of the dirt road at the southern end of Wapi Park. A walk that does NOT follow a trail to and from the Pillar Butte vent complex constitutes STOP 3 and begins and ends at this locality (STOP 3). The traverse is about 5 km long and requires about 4 hours. Water should be carried on this traverse, particularly in the hot summer months. Walk about half way from STOP 3 to the north end of the Pillar Butte vent complex. As a landmark during the traverse, walk toward the left of Pillar Butte. The hummocky relief of the lava surface is characterized by pressure ridges and local collapse depressions. Local relief changes from low over the pressure plateau areas to fairly high (10 m) near flow ridges. This form of topography is characteristic of the margins of the Wapi field, but quite different from that of the Pillar Butte vent complex (**Figure 58A**).

Continue south-southeast for about one mile to the contact between lavas of the main Wapi lava field. Observe tube and surface fed flows of the Pillar Butte vent complex. The Pillar Butte flows at this locality emanated from the northwest side of the Pillar Butte area. Tubes and channels in the flows and small pit craters are present (**Figure 58B**).

Proceed southeast to the north flank of the summit area of Pillar Butte. This area is a complex of several pit craters and vents that fed flows to the northern part of the lava field. Lava flows of the summit area consist chiefly of shelly pahoehoe that contains numerous channels that slope to the north. Shelly pahoehoe forms near vent areas from gas charged fluid lavas (**Figure 58C**).

Traverse south along the eastern margin of the summit region, noting more shelly pahoehoe and slab lava. Proceed to the eastern edge of the major pit crater complex of the Pillar Butte vent area (**Figure 58D**). The western wall of the vent area displays layered rootless flows of the summit region in cross section. In the pit craters, observe ledges of lava lakes that filled the pit craters and then subsided. Thin deposits of lapilli generated by vigorous fountaining occur along the east margin of the vent area.

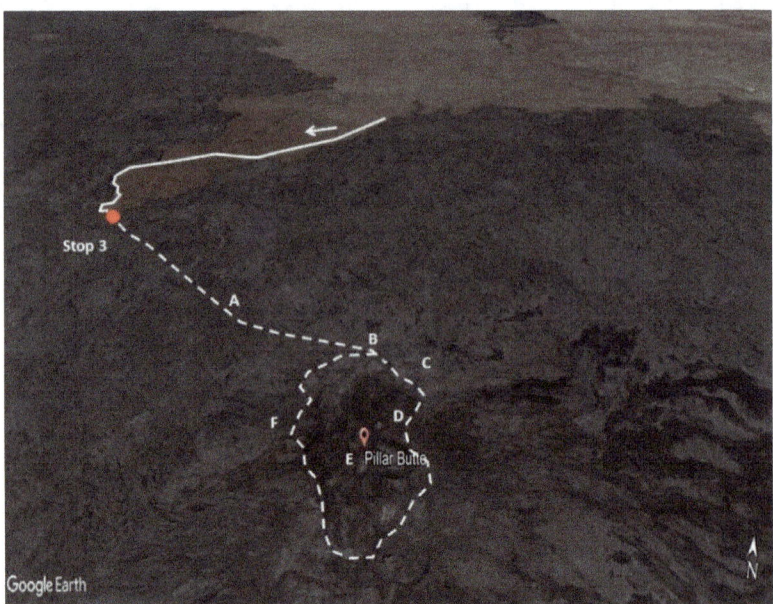

Figure 58. Stop 3 includes five observations points surrounding Pillar Butte.

Continue south to Pillar Butte, an excellent vantage point (**Figure 58E**) for the general aspects of the Wapi lava field as well as the topography and features of the eastern Snake River Plain can be viewed from here. The origin of Pillar Butte is unclear. Its relief is due either to deflation of the area around it or to inflation directly beneath it.

Proceed northwest to an area containing well-exposed east facing subsidence faults that have truncated flow channels and tubes (**Figure 58F**). The orientation of these faults parallel fissures of the Kings Bowl Rift system. From this locality, continue northwest, and then west-northwest back to STOP 3 where the vehicles are parked.

Stop 4. Split Butte

Follow the dirt road to the west side of the Wapi lava field. A dirt road leads off to the left to Split Butte behind a large crater (**Figure 59**).

Walk along the rim to observe tephra sequences, then down the inner slope to the edge of the basalt cliff. Located about one km west of the Wapi lava field, Split Butte is a maar type phreato-magmatic crater that is one of the older exposed features on the ESRP. It consists of a subcircular tephra ring 0.6 km in diameter with an inner basalt lava lake which suggests that an initial eruptive phase caused by basaltic magma interacting with groundwater was followed by an effusive phase.

Figure 59. Split Butte is the ridge in the background behind the large crater in the foreground. A dirt road passes through the center of the large crater providing access to Split Butte.

The inner lava lake rose to a depth within the ring that was higher than the surrounding lava surface. After minor overflow on the southwest flank, it subsided to leave a circular terrace like platform around the inside margin of the crater. The tephra ring is surrounded by loess-covered Quaternary ESRP basalts. The Holocene Wapi flows stopped short of lapping up on the southeast flank. The lava lake effusive eruption was probably fairly quiet as indicated by a lack of spatter, although slumping within the tephra ring resulted in a disconformable contact visible between it and the surrounding basalt lava.

Several lines of evidence points to a phreato-magmatic origin. The ash is mostly clearly palagonitized sideromelane that is typically blocky and angular with few vesicles indicating rapid quenching in a wet environment. Plastic deformation of ash layers also occurs beneath ejecta blocks, and secondary minerals such as calcite and zeolite are abundant. Pyroclastics typical of strombolian eruptions are generally absent and layers of hyaloclastite rich coarse tephra are interspersed with fine ash having accretionary lapilli. This is the final stop of the field trip.

Drive south on the dirt road to Minidoka on Idaho 24. Continue to Rupert and Hayburn about 17 mi southwest to the 1-84 Freeway. Take Interstate 84 east back to the junction of Interstate 86. Take Interstate 86 back to American Falls and Pocatello.   End of Field Trip.

Photo Gallery of the Eastern Snake River Plain and the Great Rift

Close up view of the southern most Big Crater vent in Craters of the Moon National Monument. Cinders appear below lava flows at the base (reddish brown on left, and black on right). Source: Photo by author taken in 1983 before the monument was developed.

Rounded hill on the right side is an example of a kiputka described in the text from Craters of the Moon National Monument. Source: Photographed by the author in 1983 before the monument was developed.

Pahoehoe lava flow taken in Craters of the Moon National Monument. Source: Photo taken by the author before the monument was developed in 1983.

An example of a squeeze up lava flow in Craters of the Moon National Monument. The flow is raised up above the flow underneath, pushed up from below through fractures in the older flow.

An example of an entrance to a lava tube in Craters of the Moon NM. The daylight streaming into the tube is called a skylight where the overlying roof caved into the void space. Source: National Park Service posted on the internet.

Collapse structures into void spaces created by lava tubes beneath the surface of the flow in Craters of the Moon NM. Source: Kool 96.5 radio station posted on the internet.

Juniper Mountain lava flows exposed along the flanks of the mountain. Source: BDSUSA posted on the internet.

View facing south of the Great Rift with an eruptive cone in the background. Source: Digital Atlas of Idaho posted on the internet.

Kings Bowl lies along a fissure responsible for erupting thin lava flows exposed along the walls of the bowl. Crystal Ice Cave occurs at the bottom of the structure and is flooded with ground water in this photograph. Source: Flickr posted on the internet.

Fissure eruptive structure occurring between Kings Bowl and South Grotto.

Pillar Falls exposed tuff in the canyon walls located half way between Shoshone Falls and the Perrine Memorial Bridge. Source: Johnny T. Cheng posted on the internet.

References

Champion, D.E., King, J.S., Covington, H.R. 1983. Kings Bowl and Wapi Lava Field and the Southern Part of the Great Rift. in Nash, W.P., Gurgel, K.D., Harper, G.D., 1983. Geologic Excursions in Volcanology: Eastern Snake River Plain (Idaho) and Southwestern Utah. Guidebook Part III. The Geological Society of America Rocky Mountain and Cordilleran Sections Meeting Salt Lake City, Utah. Utah Geological and Mineral Survey Special Studies 61.

Hughes, S.S., Smith, R.E., Hackett, W.R., McCurry, M., Anderson, S.R., Ferdock, G.C. Bimodal Magmatism, Basaltic Volcanic Styles, Tectonics, and Geomorphic Processes of the Eastern Snake River Plain, Idaho. Field Trip Guidebook. 1997 Annual Meeting, Salt Lake City, Utah. Geological Society of America Part I Volume 42.

Kuntz, M.A., Lefebvre, R.H., Champion, D.E., King, J.S., Covington, H.R. 1983. Holocene Basaltic volcanism along the Great Rift, Central and Eastern Snake River Plain, Idaho. Part I: Craters of the Moon lava field and the northern part of the Great Rift; Part II: Kings Bowl and Wapi lava fields and the southern part of the Great Rift in Nash, W.P., Gurgel, K.D., Harper, G.D., 1983. Geologic Excursions in Volcanology: Eastern Snake River Plain (Idaho) and Southwestern Utah. Guidebook Part III. The Geological Society of America Rocky Mountain and Cordilleran Sections Meeting Salt Lake City, Utah. Utah Geological and Mineral Survey Special Studies 61.

Other Snake River Books Available on Amazon

Book Price: $34.95; Ebook Price: $17.48.

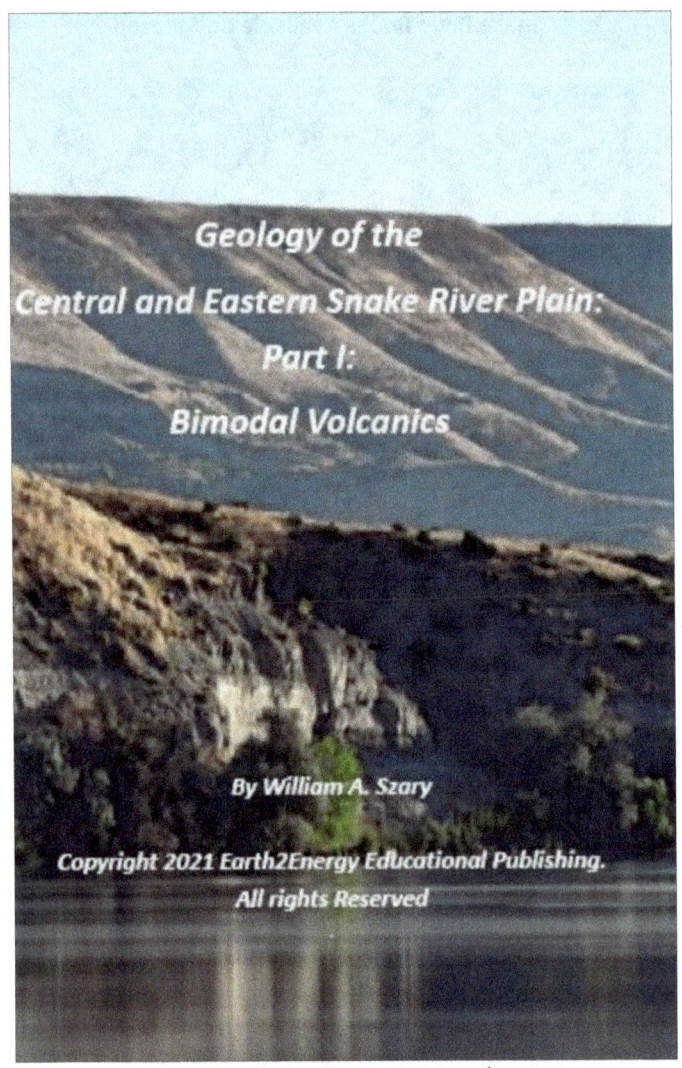

Book Price: $42.38; Ebook Price: $21.19.

www.ingramcontent.com/pod-product-compliance
Lightning Source LLC
Chambersburg PA
CBHW052339220526
45472CB00001B/487